EMP-Hardened
Radio Communications

By William T. Prepperdoc

Version 1.1

Significant portions of the text of this book were previously posted in TheSurvivalistBlog.net and in Survivalblog.com, and are used with the permission of their respective editors.

This short manual is dedicated to James Wesley, Rawles, and M.D. Creekmore, for their tireless efforts to assist ordinary citizens in becoming better prepared to thrive in difficult life circumstances.

CONTENTS

FOREWORD

The risks posed by coronal mass ejections (CMEs) from the sun and electromagnetic pulses (EMPs) are real, and their risks to society are gradually increasing year after year. In First World nations, we now carry out our lives almost inextricably tied to the power grids, long distance telecommunications, Internet-based inventory control systems, electronic banking, and with microcircuits that are increasingly embedded in a dizzying array of home and business electronics. Even our dishwashers, clothes washers and toasters now have microchips in their control systems. Our western society has unwittingly become a house of cards.

Most alarmingly, not only are microchips now diffused in a larger number of systems, but the chips themselves are growing more vulnerable, because their gate dimensions are decreasing while their transistor density is increasing, with ongoing advances in chip design microlithography processes. Around 15 years ago, most commercial microchips had gate dimensions that were measured in the range of 1 to 3 microns. But today, gates as small as 3/10ths of a micron have become commonplace. Submicron gates have become the norm. In essence: *The smaller the gate, the more vulnerable it is to being fused by fast rise time pulses created by a CME or EMP.*

The manifold risks posed by CMEs and EMPs cannot be overstated. A major pulse event would bring modern technological societies to their knees. Recovery could take years or even *decades*. A slow process of rebuilding infrastructure would have to take place all along the path from power generation facilities down the line through power distribution networks, and finally to the microcircuits in every grid-connected device in homes, offices, and factories. Even entirely off-grid homes powered by photovoltaics or microhydro power are vulnerable, if any radios are connected to antennas. (Since antennas can act as conduits for transient pulses, just as power lines and phone lines do.)

Most treatises on EMP and CME protection focus on simple Faraday cage protection for electronics that are not in regular use. That is all well and good for your *spare* radios, *spare* computers, *spare* photovoltaic charge controllers, and your *spare* inverters. That is sound advice that should not be overlooked. But what about the electronics that you *must* use on a day-to-day basis? "William T. Prepperdoc" has finally written the book that many prepared individuals have been looking for. In *EMP-Hardened Radio Communications*, he details not just the risks of CMEs and EMPs, but also precisely how to retrofit existing hardware to give it fairly robust hardening, on a modest budget. The practical specificity of his text is tremendous, and it is something rarely seen in publications outside of arcane technical circles. This is a reference for the common man.

The book that you are holding in your hands could save the lives of your loved ones. And if you are a business owner, it could mean the difference between getting your company back up and running quickly, or never recovering at all.

You are probably already aware that the sun goes through an 11-year cycle of activity. As this book goes to press, we've just experienced a very weak sunspot maximum, and we are now heading toward a sunspot minimum. Strangely, it is during sunspot *minimums* that the risk of X-

Class solar flares is greatest. (It was during such a minimum that the often-cited Carrington Event occurred.) With that fact in mind, you should have a sense of urgency in your preparations.

As a Christian, I believe that nothing happens by chance. Our lives are truly in God's hands, and we are fulfilling destinies that predate the very foundation of the Earth. Please prayerfully consider the advice given in this book, and put it into action, *soon.*

It wasn't raining when Noah built the Ark.

James Wesley, Rawles
Founder and Senior Editor of SurvivalBlog.com
December, 2016 - The Rawles Ranch, in The American Redoubt

ACKNOWLEDGMENTS

All of us are touched by large numbers of people around us whose influence enhances our lives. I'd like to especially acknowledge the following special people:

- Sam R. for introducing me not only to solar power, but also diesel automobiles!
- Duke B. for putting up with my crazy projects and being an all-around friend;
- Bud T. for his mentorship in WINLINK;
- MD Creekmore and James Wesley, Rawles for their support of all my writing & educational efforts and their encouragement to write this text;
- Jeff C. for his amazing support of my many ham radio emergency communications ideas;
- Most of all, Nancy, who always loves and supports me in my various projects no matter how big a mess I make of the house in the process.

CHAPTER 1
INTRODUCTION

For wisdom is better than jewels;
And all desirable things cannot compare with her.
Proverbs 8:11

The purpose of this book is to assist the average citizen in becoming prepared to maintain radio communications despite electromagnetic threats to electrical/electronic equipment and to the national electrical grid. More and more citizens of many nations have become aware of the EMP (electromagnetic pulse) & CME (coronal mass ejection) risks to electronics & to national electrical grids, through movies, Internet literature, and even through Ted Koppel's recent book <u>Lights Out,</u> (although Koppel focused on *cyber-warfare* risks to the electrical grid).

Cell phones *are* <u>radios</u>. ***Broadcast television*** is at its essence, a <u>radio</u>. ***Pagers*** (beepers) are <u>radios</u>. ***Cable television*** is basically a <u>radio</u> system with a long transmission line missing an antenna --- and is quite vulnerable to EMP/CME.

The controls for almost all of our infrastructure, including SCADA (supervisory control and data acquisition) computers managing fuel and energy flows throughout most nations, are <u>radio</u>-based. Airplanes are tracked and guided by <u>radio</u>; their pilots use <u>radio</u> for all communications. The same is true of boats and trains. Satellites are <u>radio</u>-based.

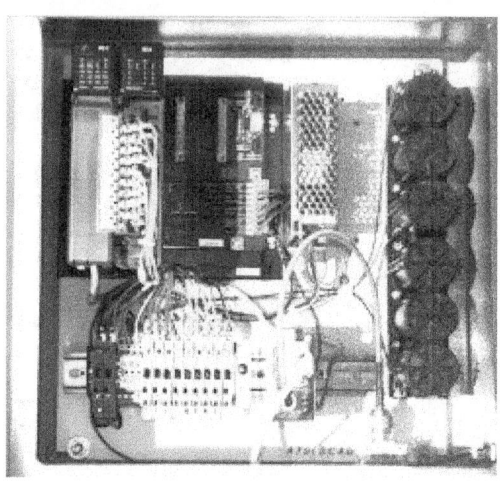

Figure 1-1 Typical SCADA computer system. [1]

Radio communications permeate, and are vital to the continued operation of modern society.

If an EMP attack or significant CME event occurs, and the population is unprepared (as it is now), the consequences are likely to be cataclysmic beyond any known historical effect, even beyond that of conventional world war. Woolsey & Pry, in a *Wall Street Journal* op-ed, stated the 2008 Commission had estimated a die-off of up to 90% [2] (although I can't find that figure within their publicly published report), due to inability to create and distribute food and water, maintain law and order, and provide medical care. Such a mortality is *far beyond* that of the world wars. Anarchy is the commonly-expected outcome, absent a miracle.

A threat with such capacity for harm should not be ignored. Thus, I write this book---because I care for people. Better preparation of the civilian population for such a threat not only includes comprehensive planning for basic foodstuffs/water/medical care/defense, but also and crucially the provision for communications that will provide precious news and potential dissemination of life-saving information throughout the population. *Or at the very least, between and throughout smaller groups and individuals who have made adequate preparations.*

The author is a practicing physician at a major university teaching medical school. Yes, that means I take care of real patients for a living, supervising resident physicians in post-MD training and practice. Prior to the career change into medicine, I was an electrical engineer, with Bachelor's and Master's degrees from prestigious schools of Electrical Engineering. More importantly, I was a very active amateur radio operator even in high school days, first building radio equipment from kits, and later designing and building radio equipment from scratch. All these experiences prepared me to understand and propose solutions to the threats posed by Electromagnetic Pulse and Coronal Mass Ejections.

Although military and nuclear experts were aware early in the Nuclear Age of the impact of Electromagnetic Pulse, this information took significantly longer to filter out into the civilian population and the other professional disciplines. Preoccupied with arduous medical training, I was blissfully unaware of these very significant risks until only a few years ago. However, a very large body of information is now readily available for study and it isn't difficult to grasp the risks and plot strategies to minimize their effects.

This book will help you do just that.

EMP-Hardened Radio Communications

CHAPTER 2
THE HISTORY OF ELECTROMAGNETIC PULSE (EMP) AND CORONAL MASS EJECTION (CME)

Without wise leadership, a nation is in trouble;
but with good counselors, there is safety.
Proverbs 11:14

The purpose of this chapter is to give the reader the history of how the vast EMP potential of nuclear weapons was finally recognized, provide some explanation for why the general public to this day is so vulnerable, and along the way introduce some of the technical details of the risks of both EMP and CME.

19th Century
Long before transistors, before nuclear weapons, before even radio, there was the *telegraph.* A simple electrical circular circuit in which a mechanically operated switch ("telegraph key") opened and closed a complete circuit, allowing current from a battery to flow through an electromagnet, causing a hinged ferromagnetic bar to rotate, strike another piece and create sound. Enlarge the wiring across vast stretches of territory, and Morse Code could be used to transmit news (or military commands or financial transactions!) across a nation. Typically, in order to save money, only one of the two required conductors in the circuit was actually made of a wire strung between telegraph poles -- the other conductor was provided by the earth below. This created a significant <u>enclosed area</u> between the two conductors as the telegraph line snaked miles and miles across the territory of the United States (and other countries).

In 1859, these telegraph circuits gave the world the first indication of the enormous electromagnetic risks posed by unexpected movement between supposedly stationary earth magnetic fields, and electrical conductors. The event is named for Richard Carrington-- an amateur astronomer who on Sept. 1, 1859, observed bright solar flares on the

5

surface of the sun, which were associated with a **coronal mass ejection** (CME) that traveled to the earth. Within hours, extremely damaging currents were being induced in telegraph wires. Morse code key contacts were spewing sparks, and actual fires resulted. The coronal mass ejection of charged particles hurtling from the sun impacted the earth's own magnetic field, shoving around the normally static lines of magnetic field from the earth. Whenever relative motion between a conductor and a magnetic field occurs, an electrical potential is created; this is the principle of the modern generator. In 1859, the "generator" had only one winding -- the telegraph wire above the earth--but it was a LONG one and the enclosed area was huge. Massive voltages were developed.

Carrington-level events are rare, but completely natural and always possible, with the world risk estimated at 12% per decade.[3] Timing and trajectory cause us to be spared often. A century may pass between unlucky targetings of the Earth. But our modern society is now orders and orders of magnitudes more vulnerable than the sturdy telegraph.

20th Century

In the first half of the 20th century, the field of Physics rapidly advanced, both in the understanding of the atom, and in the creation of electronic devices such as the vacuum tube. The radio was created in 1900, credit going to Marconi; the vacuum tube in 1906 by Lee de Forest. Nobel prizes were being won year after year by physicists who were figuring out the principles that would allow the nuclear bomb.

However, those physicists did not initially recognize the potential for vast electromagnetic effects from a nuclear explosion (and of course, the Sun is a continuous nuclear explosion). Enrico Fermi had some grasp of the possibilities, but Nobel laureate Hans Bethe misunderstood the mechanism that would allow EMP (electromagnetic pulse), and incorrectly predicted resulting electromagnetic fields that were orders of magnitude below what actually happens.

The United States was very lucky that several of our nuclear tests did not themselves bring about the first EMP disaster. In order to create the mechanism required for disastrous EMP, the nuclear explosion must occur high in the atmosphere (in an area without significant conductivity provided by existing atoms and ions) but exposed to the earth's magnetic field.

We dodged the bullet at the world's first man-made nuclear explosion[a] carried out in 1945 at Los Alamos --- carried out just above ground level instead of high in atmosphere (where it would have caused a self-inflicted EMP event).

Both bombs dropped over Japan were fused to explode below 2000 feet altitude, for maximum ground devastation. Again an EMP was avoided. [4]

It was not until 1962 that the United States (and Russia) carried out **high altitude** nuclear explosions. Testifying before the House Armed Services Committee (Oct 7, 1999, U.S. Congress), physicist Dr. Lowell Wood observed

> *"Most fortunately, these tests took place over Johnston Island in the mid-Pacific rather than the Nevada Test Site, or electromagnetic pulse would still be indelibly imprinted in the minds of the citizenry of the western U.S. as well as in the history books. As it was, significant damage was done to both civilian and military electrical systems throughout the Hawaiian Islands, over 800 miles away from ground zero. The origin and nature of this damage was successfully obscured at the time -- aided by its mysterious character and the essentially incredible truth."*

This test, code-named Starfish Prime, occurred at 59 minutes, 51 seconds before midnight, on the night of July 8 1962 (Honolulu time). A 1.44 megaton W49 thermonuclear warhead was detonated at 400 km altitude. The flash was visible 2000 km away in the Kwajalein Island sky. The sky glowed green over the south Pacific for a second. High frequency radio propagation was damaged for minutes to hours depending on the frequency & propagation path. Although an electromagnetic pulse (**HEMP**: high-altitude EMP) effect was expected, calculations based on Bethe's mistaken understanding severely underestimated both the speed and the impact, and thus most measuring instruments simply went completely & almost instantaneously off scale! Electronic items, both civilian and military, were destroyed in Hawaii, 897 miles from the explosion. About 300 city streetlamps were destroyed (admittedly a small portion of the Hawaiian streetlamps) and quite a few automobile ignition systems were destroyed. The damage was minuscule

---

[a] It is believed that natural, long-lived, fission reactions have occurred in the distant past at more than one known high concentration deposition of uranium within the earth's crust.

<u>compared to what might happen today</u>, for several reasons:

1) the small size of the Hawaiian islands meant electric power wiring systems could not be very long;

2) Starfish Prime was not optimally designed for EMP production; simpler fission nuclear explosions of much lower yield turn out to be better EMP weapons, much more efficiently producing EMP, particularly the fast-rise E1 component; [5]

3) the magnetic field at the location produced only a fraction of the EMP strength expected in a strike against the U.S.; and

4) **there were almost no transistorized electronic equipment in use in 1962.** Far more rugged vacuum tubes ran almost all electronic equipment of that day.

In the same year, 1962, the Soviet Union carried out multiple nuclear tests including high altitude tests. In these tests, the following damages were observed:

- puncture and temporary disconnection of overhead high voltage transmission lines
- spark gap breakdown & safety devices caught fire, on overhead signals lines 600 km away from ground zero
- malfunction of radiolocation equipment 1000 km away
- power supply breakdown of signal cable line, 600 km distant
- damage to diesel generators discovered "later"[6]

In the decade that followed those high-altitude experiments by the major powers, the physicists got their understanding of the EMP potential straightened out. Briefly and superficially, there are three electromagnetic impacts of a high altitude nuclear explosion (HEMP):

1. E1 -- an extraordinarily fast burst of wide-band radio frequency energy that does most of the immediate damage to electronic circuits

2. E2 -- a slightly delayed burst of wide band radio energy, not as destructive as the E1, and more similar to lightning.

3. E3 -- a late and slow impact caused by the shoving aside of the earth's magnetic field by the nuclear blast, resulting in high currents in the same types of wiring system impacted by the Carrington event --- that begins half a minute or more after the blast and lasts half a minute or so.

As the physicists got their understanding of the EMP nuclear attack risks, there was a move to protect military systems from destruction by such a single blast. Concurrently there were feeble efforts to increase the level of protection of electronic equipment in the hands of the civilian population, most of which was switching from the tough vacuum tube to the fragile transistor and later, the integrated circuit.[a]

In those days, amateur radio operators were considered a valuable line of last-resort radio communications in the general population. In a series of four papers published in *QST*, the national journal of the amateur radio operators of the United States, beginning in August 1986, Dennis Bodson, an amateur radio operator (W4PWF) and the Acting Assistant Manager, Technology and Standards, National Communications System, of the United States federal government carefully explained the radio threat posed by EMP. His articles are well-written, very detailed and also very practical, and extremely valuable to readers today. This publicly-disseminated information was certainly approved by the U.S. Government, likely in an attempt to harden the "emergency communications structure of last resort"---amateur radio operators--- in the United States.

Bodson not only explained the mechanism of EMP damage, but also provided detailed examples of protective techniques -- and even provided the results of testing of those techniques in EMP simulators, proving that radio systems could indeed be hardened significantly against EMP attack.[7]

On March 13, 1989, the earth experienced a modest version of the Carrington-type of CME solar-induced electrical event involving Quebec, the northeastern United States and the United Kingdom. The 2008 Congressional Report recorded:

> *"On this day, several major impacts occurred to the power grids in North America and the United Kingdom. This included the complete blackout of the Hydro-Quebec power system and damage to two 400/275 kV autotransformers in southern England. In addition, at the Salem nuclear power plant in New Jersey, a 1200 MVA, 500 kV*

---

[a]   Vacuum tubes exposed to overvoltages may arc internally from one metal structure to another; those that are not filled with argon or another inert gas may spontaneously extinguish the arc at the end of the sub-microsecond E1 event, and continue to operate. Transistors and integrated circuits (now made incredibly tiny) contain P-N junctions (or insulated gates) and are typically irrevocably destroyed by any overvoltage.

transformer was damaged beyond repair when portions of its structure failed due to thermal stress. The failure was caused by stray magnetic flux impinging on the transformer core. **Fortunately, a replacement transformer was readily available; otherwise the plant would have been down for a year, which is the normal delivery time for larger power transformers.** [emphasis added] *The two autotransformers in southern England were also damaged from stray flux that produced hot spots, which caused significant gassing from the breakdown of the insulating oil.*

The blackout of the Hydro-Quebec system was caused when seven static voltage-amps reactive (VAR) compensators (SVC) tripped and shut down due to increased levels of harmonics on the power lines. The loss of the seven SVCs led to voltage depression and frequency increase on the system, which caused part of the Quebec grid to collapse. Soon afterwards, the rest of the grid collapsed because of the abrupt loss of load and generation. The blackout took less than 90 seconds to occur after the first SVC tripped. About 6 million people were left without power for several hours and, even 9 hours later, there were still 1 million people without power."[8]

Figure 2-1. Destroyed transformer caused by the 1989 Geomagnetic Storm.[9]

(Just recently, a far more potentially destructive solar flare occurred, but luckily the Earth was not positioned on its orbit around the sun to receive the energy, or we might have gotten a taste of life without electricity.[10])

Unfortunately, the nation was preoccupied with other world issues for much of the latter quarter of the 20th century, including the collapse of the Soviet Union, the development of the Internet, and the dot com revolution. With our prime adversary crumbling, a score of years elapsed since the nation's last significant war experience (the disastrous Vietnam war) , and the stock market of the 1990's soaring, the average American wasn't putting too much thought into EMP vulnerability. Y2K hysteria came and went without any significant problem. What could go wrong?

21st Century

Then the dot com stock market crash had its first rumblings, and the fateful day of September 11, 2001 arrived, with simultaneous attacks on several American cities, and the first complete shutdown of the entire American airspace.

Just prior to the 9/11 attack Congress had commissioned a report on EMP vulnerabilities. That report was released in full in 2004[11] as the nation was embroiled in the War on Terror and more attention began to be focused on the national EMP vulnerabilities. A follow-on report was commissioned. The stark potential for catastrophe in one after another computerized segment of the national power grid, transportation systems, financial systems -- on and on --- was fully described in the resulting 2008 report. [12]

North Korea developed a nuclear bomb (first test Oct. 9 2006[13])-- and then several. Iran began spinning uranium to concentrate fissionable U-235. North Korea began developing--and testing-- missiles quite sufficient to launch a nuclear weapon high into the atmosphere.

The Commission writing the 2008 report found that military and scientific literature from Taiwan, Israel, Egypt, India Pakistan, Iran and North Korea demonstrated knowledge of EMP and its military usage. Russian Duma Leaders in 1999 had even boasted to a U.S. Congressional Delegation that a single ballistic missile detonated at high altitude over the United States would provide devastating EMP retaliation toward the U.S. if desired.

But the United States was awakening to a new threat: terrorist attacks by small but terribly effective non-national actors.

The fictional book One Second After, based on a fictional EMP attack, was published in March, 2009, and reached #11 in the *New York*

Times Best Seller list. With the Afghanistan, Iraq, Libyan and other wars not going well, the nation began to recognize the real possibility of crippling attacks as well as naturally-occurring solar-induced power system failure. The possibility of a barge-launched nuclear missile inside the U.S. ICBM defenses, resulting in an EMP catastrophe, became more realistic. In 2014 the *Wall Street Journal* published an op-ed by a former head of the CIA highlighting the risks.[14] "Prepper" and survivalist online Internet literature became replete with information on the devastation that might result, and government experts predicted a die-off of up to 90% of the U.S. population. With the increased national dependence on the Internet and computerization, the consequences of loss of electrical power and telecommunications became clearer to the public.

In April, 2015 the U.S. NORAD central command announced a $700 million contract with Raytheon to provide EMP protection upgrades allowing NORAD to move ***back inside EMP-safe Cheyenne Mountain***, literally putting their computers inside of a mountain. [15]

A small but growing minority of the American population paid attention and the modern "prepper" or "survivalist" movement was born. Although there were several disaster scenarios of great concern to those citizens (Rawles' famous novel **Patriots** revolved around a debt-created financial catastrophe), EMP/CME events figured highly in most "prepper" plans. Even *National Geographic* got involved, with the creation of a docudrama "American Blackout" that vividly portrayed the likely consequences of a wide-scale loss of the power grid (in that movie, due to cyber warfare.) Famous TV anchorman Ted Koppel wrote Lights Out on the same theme. Preppers created alternative power generation systems, studied protection against EMP, and learned the homesteading techniques of self-preparedness

EMP is now so well-known in our society that there is a cottage industry building locally-effective (non-nuclear!) EMP generators based on high power RF emissions that actually work.[16]

Nevertheless only pitifully small measures were taken to actually deal with the risks. Despite relatively inexpensive proposed solutions to address the E3 or CME risks to the national power grid, estimated at $2 Billion, almost nothing was done.

Cyber warfare --- another significant risk -- was highlighted when the

United States itself employed a software attack against Iran to delay the creation of sufficient fissionable material for nuclear weapons, and again later when Russia apparently downed a significant portion of the Ukranian power grid as a political "warning."

Political stalemate in the Obama presidency years continued, and the United States enters the Trump presidency still woefully unprepared for both EMP attack and naturally occurring Carrington-type CME events. Private electrical generation and distribution companies appear to impede the process of protecting the grid; there has been no significant national legislation, and only the State of Maine has taken truly effective action. [17]

The true impact of a high altitude EMP attack on the United States can only be estimated, as it has been by two Congressionally-mandated reports (2004 and 2008).   **It will not be known for certain until it occurs.**

INCREASING VULNERABILITY
Thus, since the tests of 1962 we have developed a far greater dependence on electronic systems, while building them out of ever-more-minute semiconductors that have far greater vulnerability, than the rugged vacuum-tube and relay-based systems of that day. While it is true that "electrostatic discharge" protections sufficient to stop static electrical failures of early transistor & integrated circuits are now commonplace, these are not expected to protect communications systems against EMP. An actual EMP attack can result in **tens of thousands of amperes** of current pulses in systems connected to wiring of any significant length, and voltages into the hundreds of thousands of volts; a CME event easily burns out enormous transformers with kilo-ampere core-saturating direct current shifts. Instead of growing more resilient against this threat, our electronic systems are becoming more and more *vulnerable.* As James Rawles noted in the Foreword, advanced semiconductor manufacturing techniques are now making the transistors of the integrated circuits on which our society depends, ever smaller. An atom is in the range of 5 Angstroms in size. By the year 2010, gate oxide thicknesses in MOS (metal oxide semiconductor) devices had reached 10 Angstroms.[18] The number of transistors per microprocessor is almost astronomical, each one now so microscopic that only a very small amount of energy is necessary for its destruction.

If you care for yourself and your loved ones, take advantage of the protection techniques described in this book.

EMP-Hardened Radio Communications

CHAPTER 3
CHARACTERISTICS OF ELECTROMAGNETIC PULSE & CORONAL MASS EJECTIONS -- HOW THEY DAMAGE COMMUNICATIONS EQUIPMENT

The mind of the intelligent seeks knowledge,
But the mouth of fools feeds on folly.
Proverbs 15: 14

In order to provide the best protection against EMP and CME, the cause and characteristics of the electromagnetic hazards must be understood. The EMP phenomenon includes events known as E1, E2, and E3; the CME solar-induced phenomenon is similar to a very long-lasting E3.

E1

So-called "prompt gamma" rays (photons of extraordinarily high frequency & energy) constitute a small portion of the energy output of a nuclear explosion. The percentage of total weapon yield that goes into the prompt gamma emissions is small for fission weapons, and usually even smaller for thermonuclear (hydrogen fusion) weapons. Design of the weapon can impact the percentage of energy released as gamma radiation. An EMP-weapon would be designed for a higher prompt gamma yield.

The gamma rays are produced extremely quickly -- beginning immediately and lasting only a few nanoseconds to fractions of a microsecond (far briefer than comparatively long-lasting lightning bolts, lasting hundreds of milliseconds). Striking molecules of the upper atmosphere, they knock loose electrons, which then join in a semi-coherent flow of electrons if the conductivity of the region is sufficiently low so that the electrons are not reabsorbed too rapidly (hence the requirement for high altitude explosion). A flow of electrons constitutes an electrical current. An electrical current flowing within a magnetic field creates an electrical field. The extremely fast onset in the prompt gamma emission results in this field having a very broad frequency spectrum. Although the strength of the field is not homogeneous in all directions (and complicated "smile" patterns demonstrating higher intensity just south of explosion, can be found in reference works, see Figure 3-1), **the extraordinarily strong and brief electromagnetic wave basically blasts downward to the earth much like the electronic flash of a camera, exposing all portions of the earth that are accessible by any straight line from the point of explosion.** A high enough explosion can impact virtually all of the lower 48 states of the United States.

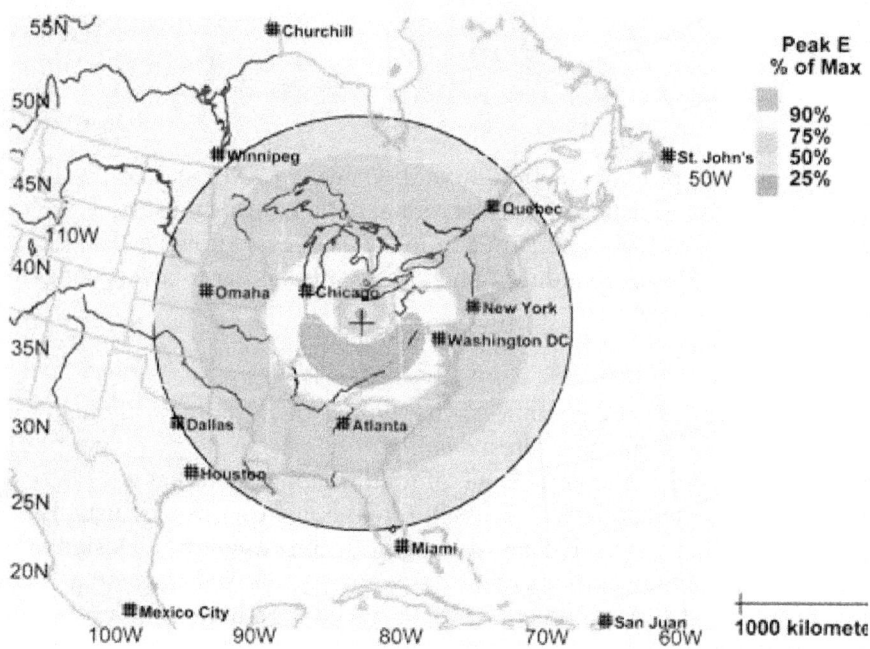

Figure 3-1 Illustration of widespread E1 field strength results from a single explosion.[19]

TABLE 3-1 Rough percentages of power distribution between different products of nuclear explosion. [20]

Type of Release	% of total power
Xray Photons	70%
Kinetic Energy	20%
Neutrons	1%
Gamma Ray Photons (genesis of E1)	0.1%

The field strength of this fractions-of-a-microsecond electromagnetic burst of radio frequency energy is phenomenally high -- up to 50 kV/meter. Thankfully, there are physical limitations that prevent it from reaching levels much over this incredible level.

The implication of the incredible field strength (orders of magnitude higher than any experienced in normal life) is that any wire (e.g., an antenna) longer than a foot or so may develop many kilovolts (possibly MANY kilovolts) of electrical potential. Depending on the impedance of the wire's connected circuitry, many kiloamperes of current may flow!

It becomes more clear why fuses and signal wires were damaged in 1962 high altitude tests, and why insulation was damaged in generators. In the tests reported in *QST* articles, it was determined that half-inch-thick RG-8 coaxial cable voltage limits were easily exceeded, resulting in internal arcing within the cable (which serendipitously placed a limit on the kilovolts that reached the vulnerable radio equipment and protective systems).

Despite the vast power of the E1 signal and its resultant capacity to destroy fragile circuitry, its frequency spectrum is not infinite. The radio frequency field is strongest below 10 MHz, weaker above 100 MHz, and relatively weaker still beyond 1 GHz. This is shown in the graph of Figure 3-2. (Note the logarithmic relative field strength scale...the power really declines above 10 MHz.)

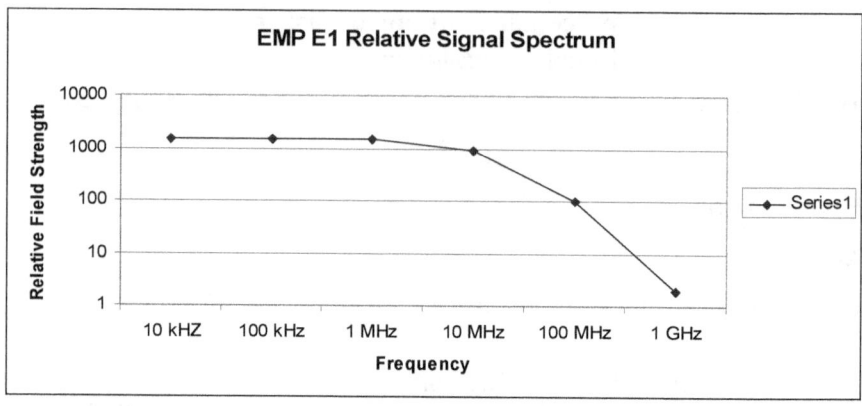

Figure 3-2: Estimated Relative Field Strength of E1, versus Frequency

This means that the E1 wave is basically an **extraordinarily strong version of radio frequency interference** (RFI) at fairly commonly-used frequencies. Ham radio operators often have dealt with RFI and can adapt their strategies to deal with this incredibly powerful version.

E2

The E2 wave occurs micro- to milliseconds after the intense E1 wave, and results from scattered gamma rays and other ejected nuclear particles, creating a somewhat lower field strength electromagnetic wave. The protection systems that are built to guard against the E1 wave will be more adequate for protection against the weaker E2 wave.

E3

The E3 wave bears more resemblance to the Carrington-type Coronal Mass Ejection disruptions. It is due to a shoving-aside of the earth's magnetic field that occurs within one- to a few hundreds of seconds after the explosion.

Whereas the E1/E2 damage is created by a fast electrical current flowing within the magnetic field of the earth; the E3 damage is created by the movement of the earth's field itself as it is shoved aside. Just as the movement of a current within a magnetic field creates an electromagnetic wave; the movement of the magnetic field encompassing earth-based conductors will create an electric potential on those conductors.

In this case, the strength of the resulting electrical potential/current is proportional to the rate of change of the magnetic field and the surface area enclosed by the circuit through which the changing magnetic field passes.

Electrical power lines traversing large distances on the earth's surface to deliver electrical power are thus uniquely vulnerable to the E3 wave, because of the large areas enclosed by their wires.

The E3 wave has the capacity to create many-seconds-long relatively constant *quasi-direct currents* of thousands of amperes within these electrical power lines --- which are part of Alternating Current circuits connected to step-up and step-down power transformers built expressly for *alternating* currents. Impressing a quasi-direct-current waveform of thousands of amperes on those transformers can easily move their magnetic fields into the saturation zones of their cores, resulting in dramatically increased losses, excessive heat, and resultant damage and fires at those transformers.[21]

The 2008 <u>Report of the Commission to Assess the Threat to the United States from Electromagnetic Pulse (EMP) Attack</u> includes (p. 43) this chilling assessment of the E3 risks of a high-altitude EMP as compared to our previous experience in the 1989 CME event:

> *"Geomagnetic storms represent an approximation to an E3-induced voltage effect. The experience to date is of events that may be orders of magnitude smaller in scope and less severe than that expected from an EMP"*

IMPLICATIONS

E1/E2
The E1/E2 impulses can be thought of as extremely powerful radio waves. Any wire longer than a few feet may act like an antenna and pick up an extremely fast and high voltage signal; any semiconductor that is downstream and not protected in some way may be destroyed. This means that radio receivers, transmitters, transceivers may be destroyed. Solid state speed controls on power tools may be destroyed. There is some chance that microprocessor controlled appliances in the

kitchen will be damaged, although metal cabinets may offer some shielding to some appliances. Lighting dimmers may be destroyed (connected to very long house wiring!). Televisions with antennas, coax feeds, or power lines may be destroyed. Telephone handset bases may be destroyed, while self-contained handsets free of their base may survive due to their small size ("aperture") and built-in electrostatic-discharge protection.. There is some chance that digital safe locks may be damaged. House intercoms are unlikely to survive. Wired burglar alarms are likely to be destroyed. Computers connected to wired (as opposed to fiber or Wifi) networks, power plugs etc., may be destroyed. Portable cell phones may survive, but the towers they attempt to reach may be destroyed or deprived of power. There is some chance that the automatic voltage regulator within home generators will be damaged. While some automobiles may still function, in areas of higher signal strength, a larger percentage of automobiles may be damaged. Vehicles of all types (boats, aircraft, etc) that are damaged while in movement may crash.

Without the kind of advanced hardening explained in this book, any transistorized electronic device of significant size connected to any wire of any significant length, is likely to be destroyed.

E3

The E3 impulse is expected to do dramatic damage to interconnection transformers in the nation's power grid. Congressional efforts to provide protection (such as high-power low-resistance additions to neutral-ground connections) have so far failed except for Maine. Local neighborhood transformers may be damaged. Large swaths of the nation will lose electrical power, and restitution may be months to years in coming. Items connected to very long wires that may have significant enclosed area between them (fence wires, for example) at the home may be damaged. Because of the failing of the power grid, unusual voltages may exist on house wiring between the time of initial damage to the time of complete failure. Both high and low voltages may possibly exist on household wiring, and higher voltages may damage electronic power supplies and circuitry that is connected and survived the E1 impulse. **There are ways to protect from these effects, as will be discussed in the following chapters.**

CHAPTER 4
PROTECTING RADIO SYSTEMS
AGAINST E1 / E2

A prudent man sees evil and hides himself;
the naive proceed and pay the penalty.
Proverbs 27:12

This chapter builds on the previous material on the makeup of the E1 (and E2) damaging waves from a high-altitude nuclear explosion, and discusses how to protect radio transmitters/receivers from that threat.

The peak E1 field strength is immense, on the order of 50 kilovolts per meter, and covers a very broad frequency spectrum, from very low frequencies, past 100 MHz; but evanescent, peaking in a few nanoseconds, and over in a mere microsecond or so The equipment survival problem posed by the E1 wave is the instantaneous destruction (by overvoltage) of transistors, diodes, integrated circuits, FETs and microprocessors by the immense voltages.   Millions of volts and thousands of amperes may flow on ordinary shortwave or ham radio antennas and feedlines from the E1 wave, for an instant. Handheld VHF/UHF walkie-talkies with antennas only a few inches long won't be hit with nearly the same level of damaging power that will assault high-frequency (HF) rigs with scores of feet of antennas and feedlines.   This chapter concerns those vulnerable high frequency or VHF stations with antennas more than 2 feet long. Of course, these are the very stations useful for getting interstate, national and international news and messages flowing.

Faraday Cages?   To protect against E1 EMP, many plan to simply store vulnerable equipment inside an effective Faraday shield -- a good quality metal trashcan with a tight-fitting lid is suggested, possibly with aluminized tape covering gaps in the lid seal.   "Nested" Faraday shields, such as wrapping aluminum foil around the sensitive radio, and placing it within a cardboard box inside the trashcan, are also often suggested, but may not be necessary if the outer shell's gaps are properly dealt with.[22]

The inadequacy of the Faraday solution is twofold: (1) it cannot protect equipment that is IN USE at the time of the attack, and (2) it basically means you don't dare USE your surviving equipment, possibly for *years* after the initial attack: "What if the enemy fires a 2nd or 3rd salvo?" one would reasonably worry. What good is equipment that remains vulnerable and may be destroyed if put to use? There is therefore a significant need to have a level of protection available for up-and-running, connected communications equipment, particular high frequency ham radio equipment, while possibly also storing spares.

[In fact, an adversary is likely to delay 2nd or 3rd EMP-explosions by at least a short period of time, because the enhanced conductivity at high altitudes due to the fact that the 1st nuclear blast will substantially reduce the effectiveness of subsequent E1 generation until the conductivity subsides.[23]]

The good news is that using simple commercially-available surge devices, this chapter will explain how a reasonable protection strategy can be implemented at a cost in the range of $20! **Part One** will handle the protection of vacuum-tube type radios, and **Part Two** will tackle the somewhat more difficult situation of transistorized HF radios (including CB). One of the great advantages of ham radio, is that the licensed users are allowed to build and modify their own stations, and this is one situation where that really pays off.

E1 EMP has characteristics of both extremely high power electromagnetic (radio) interference, as well as a very nearby or direct lightning strike. (Lightning also generates enormous radio waves, and is a primary cause of background noise on HF frequencies.) Hardening techniques therefore are a combination of filtering, shielding and hefty over-voltage circuit protection devices. A side benefit will be an increased level of protection against nearby lightning strikes.

In a widely-disseminated 4-part series of articles published in *QST* during 1986, Dennis Bodson of the Office of Technology and Standards, National Communications System, explained the problem in detail and tested possible solutions. [24] [25] Using his recommended solutions, vacuum tube radio equipment survived simulated EMP attack without difficulty, and solid-state radios often did as well, even when connected to antennas. The purpose of this article is to explain how one can easily implement those solutions with HF communications equipment and commercially available surge protection devices available right now.

Solutions for Connected Tube-Type Equipment: No sense tempting fate: when the equipment is *not* in use, a shorting antenna switch near to the entry of the transmission feedline to the residence can be utilized to conveniently disconnect the equipment from the antenna, and further to automatically short out the antenna transmission line. An example of such a switch (there are many) is the OPEK Model CX-201 2-output coaxial switch.[26] When an antenna is deselected, this switch mechanically grounds the feedline from that antenna, which would knock down the EMP E1 reaching your equipment by many orders of magnitude (manufacturer claims 50dB).

Secondly, simply using coaxial cable as part of the feedline system provides some significant amount of protection, as it will arc over with a few kilovolts (depending on the type of cable utilized), somewhat limiting the voltage. (It might be wise to keep some spare coax cable in your Faraday storage compartment.)

Voltage clamp devices are the next--and possibly most important-- protection to employ. But how does one choose the clamp voltage level? If you are going to transmit, you must ensure that your protection system will not short out your own transmissions (potentially damaging your transmitter)! The transmitter's output RF voltage is dependent on the standing wave ratio (lower SWR means lower voltage), the feedline impedance (for coax, usually 50 or 75 ohms), and transmitter power. For reference, a 100-watt transmitter output with a 50-ohm transmission line and a well-matched antenna with a 1.5:1 SWR will generate about 110 volts of RF AC in its feedline while transmitting. Obviously, a protection system that shorted out any voltage over 100 volts won't work well in this instance. The clamping voltage needs to be higher, particularly if you may have worse SWR or a linear amplifier.

A very simple solution that accommodates and reasonably protects most higher-power vacuum-tube based ham radio stations, and which indeed performed very well in Bodson's tests, was to take two 350-volt small gas-discharge surge arrestors (BI-A350; see Figure 4-1; specification sheet Ref. [27]), wire them in series, and then connect the resulting package between the center wire of the coax transmission line and the shield, with as short a wire length as practical. Bodson suggested building the gas-discharge assembly in a shielded cap on one port of a

SO-239 T-connector -- and showed that this worked very well both for protecting solid state transceivers of the 1980's and also for transmission and reception at HF/VHF/UHF frequencies. These gas-discharge tubes can carry surge currents from 5-20 *thousand* amperes. (The device lifetime is shorter with higher current surges.) They are under $4 each from multiple suppliers.[28] With two in series, the breakdown voltage will be in the neighborhood of 700 volts.

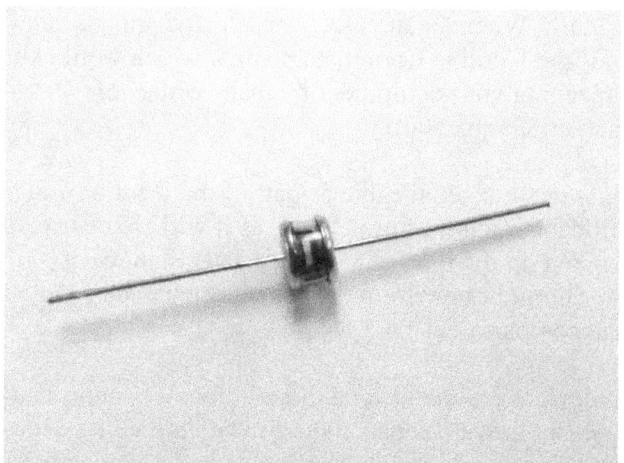

Figure 4-1. 350-volt clamping gas discharge surge arrestor.

Installing this versatile protection device requires only a small amount of wiring / soldering. Figure 4-2 shows these surge arrestors applied to the center conductor of a popular SWR measurement device. This solution is quite adequate for HF systems, and the author has used it in numerous 2-meter applications as well with success (the cheap SWR meter does not accurately measure SWR, it merely demonstrates output power in emergency communications equipment.) The SWR meter would then be connected to the transmitter side of the shorting antenna switch described above.

Figure 4-2. Two gas-discharge voltage clamping devices in series connected between center conductor of popular standing wave ratio measurement device, and the shield/ground. A small bit of insulation slid over the wiring avoids contact with the metallic frame, and after this photo was taken, insulating plastic tape was used to cover the devices prior to reinstalling the cover.

Figure 4-3 shows these soldered into a commercial antenna matching tuner. They should be installed on the side of the circuitry that has the 50-ohm impedance (closest to the transmitter, not the antenna side), rather than the possibly high-impedance antenna side, where normal voltages may be much higher.

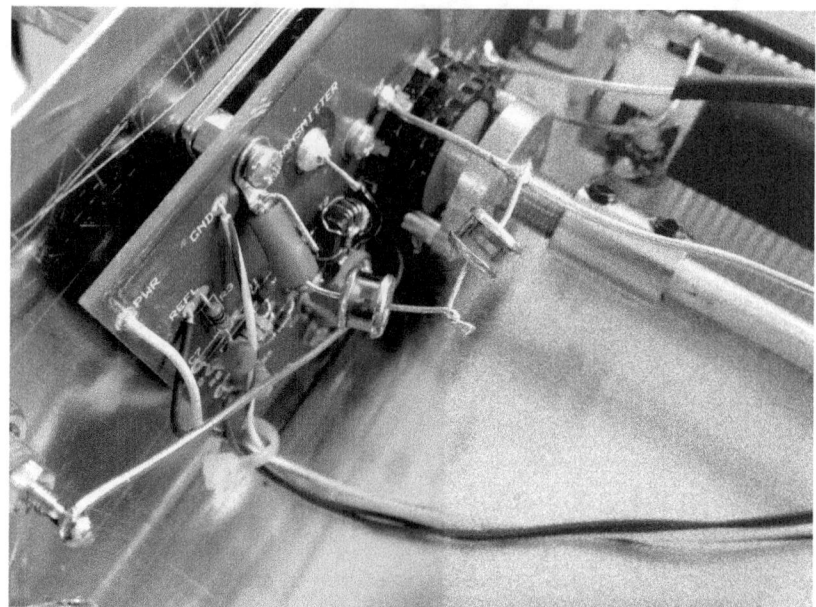

Figure 4-3. Two gas-discharge voltage clamping devices in series connected to the center conductor at the "transmitter end" of the circuitry of a popular antenna matching device, and then (red crimp terminal) connected to chassis "ground." This location was chosen so as to provide some protection to the SWR measurement portion of the antenna matching tuner.

For those not ready to modify such equipment, Bodson described a simple technique of adding these gas-discharge surge arrestors to the transmission line by using a common "UHF - T" coax connector, and his article gives good instructions (and a diagram) for doing so.

This level of protection was adequate in Bodson's tests to successfully protect vacuum-tube based ham radio equipment (receivers & transmitters) from simulated E1 EMP-induced voltages.

LIMITATIONS OF GAS-DISCHARGE ARRESTORS

A commonly-voiced reservation about gas-discharge tubes as EMP protection must be addressed at this point. As Han[29] (and no doubt many others) have pointed out, the onset of arc-formation in a gas-discharge tube is too slow to capture the leading upstroke edge of a full-bandwidth EMP E1 damaging signal.   With an EMP rise time in the range of a nanosecond, damaging voltages can easily be reached before the gas-discharge tube is able to begin conducting.   How then can they possibly be used in EMP protection circuitry, and how did Bodson demonstrate their success in protecting equipment that he clearly demonstrated was easily damaged without their protection?

The answer to this perplexing contradiction requires that the reader understand that the goal of engineering is NOT to build the perfect solution to all the world's problems---it is only to build the simplest, cheapest solution <u>that works for the problem at hand</u>. There is a big difference.

Bodson was not designing for, or testing equipment that had Gigahertz bandwidth performance capabilities; he was working with amateur radio transceivers, some of which would work in the range of 3.5-30 MHz, others between 144-148 MHz, and others in the 440MHz band.   The bandwidth of the transmitter and receiver circuitry employed would have ranged from one megahertz or so, to less than 10 MHz.

Jeng et. al., have demonstrated that equipment with a huge bandwidth (300 MHz) at a center frequency of 2.5 GHz indeed cannot be properly protected by the comparatively slow protection onset of a gas discharge tube -- and that this protection could be significantly improved by the addition of a filtering system that modified the EMP waveform upstroke.[30]

The same idea is what saves ordinary amateur radio equipment from EMP damage with ordinary gas-discharge tube protection (despite the inadequately fast protection onset by the gas-discharge tube).   The antenna tuners, receiver input filters, and power amplifier output impedance matching systems all have fairly limited bandwidth. *Indeed, it is exceedingly difficult to even create radio equipment with bandwidths of 300 MHz and operating frequencies in the Gigahertz range!*   Anyone familiar with the Fast Fourier Transform understands that the extremely fast risetime of the EMP E1 signal is formed in large part by the highest-frequency components of the E1 signal --- those in the hundreds of MHz and up to a GHz.

When you filter out those extremely high-frequency components (possibly by tens to scores of decibels) by the ordinary antenna matching filters of a ham radio system's limited bandwidth--- you eliminate that sharp upstroke and you slow down the risetime of the wave that actually reaches the equipment, apparently long enough to allow the gas-discharge tube protection circuitry to be effective.

That very likely explains Bodson's success in real-life tests.

Although there are metal oxide varistors (MOV) and other devices that in some cases have faster onset of protection than gas discharge tubes, these usually have such a **high capacitance** that they completely detune and render useless the very transceivers they are intended to protect when connected to transmission lines. Bodson found exactly that problem with some MOV protection schemes when inserted in antenna feedlines. **The very low capacitance of the gas-discharge tube leaves it as the only currently workable EMP solution for most antenna-side EMP protection systems.** (MOV's and zener-based systems are preferable on the power line side, as discussed in a later chapter.)

And indeed, **there is an entire industry that markets gas-discharge-tube-based EMP protection systems**,[31] and another that markets filtering systems for power supplies that rely even more heavily on simply filtering out the fast EMP E1 signal. [32] *You can easily purchase commercially made gas-discharge tube based feedline protection systems on the market today, already fitted with antenna connectors of your choice, if you don't want to simply build them yourself.*

The same type gas-discharge tube technology is utilized in many higher-quality lightning arrestor products intended for coaxial transmission lines.

How do we then make best use of a gas-discharge tube protection system?

1. If coaxial cable is used as the transmission line, it provides not only an inherent arc-over voltage (in Bodson's tests, around 5 kV) but also, its increasing lossiness at higher frequencies acts as a filter in itself to slow down the rise time of the signal that reaches the transceiver. **Moral: use coax, and use cheaper, smaller, more lossy coax where**

acceptable, as it provides you more EMP protection.
(And keep spare coax in your Faraday storage.)

2. An antenna matching system placed between the antenna and the gas-discharge protection device + transceiver will also tend to add another layer of protection. Typical current MFJ antenna tuners often use the T-network configuration. Although this is somewhat a "high-pass" network, at the least it tends to short out the lowest frequency components of the E1 signal, removing a lot of power and thus voltage from the signal. Much older amateur radio antenna tuners often used resonant parallel tuned circuits with coupling coils to the antenna, or a pi-network impedance matching network, and many current automated tuners use an L network, which may often end up with a series inductor and a parallel capacitor (a low pass filter). All three of these will tend to reduce the higher frequency components, thus slowing down the risetime of the damaging signal. **Moral: lean towards using an antenna tuner where possible.** (And recognize it may be damaged if not a completely manual device.)

3. Most older vacuum tube equipment uses the pi network (series inductance with capacitors to ground at each end of the inductor, mimicing the greek letter pi) to match 50 ohms into the much higher output impedance of the vacuum tube --- and the pi network is again a low-pass filter that slows down the risetime, so that the voltage actually reaching the first solid state device may not become damaging before the gas-discharge tube has fired. (And besides, vacuum tube equipment is inherently robust against EMP.) **Moral: vacuum tube equipment has continued usefulness.** (And I use it successfully for the most advanced digital communications techniques on a routine basis.)

4. Newer solid state equipment has to match a 50 ohm antenna into a much lower power transistor output impedance --- but may well use the very helpful pi-network for that job as well (just positions of large and small capacitors are reversed) and add to protection. **Moral: take a look at your equipment's schematic and see if it seems to have a pi-network interfacing transmission line and circuitry!**

PART TWO: Transistor / QRP Equipment

The solutions shown so far may not adequately protect computer-based software-defined radios, hybrid transistorized-vacuum tube radios, or fully solid-state QRP or low-power stations or transistorized shortwave radios. These radios will need to use protection that clamps at a much lower voltage than the 700 volts of two BA-350's in series! While transistor transmitters may use power amplifier transistors that may have breakdown voltages in the scores to hundreds of volts, the input circuitry of many QRP transistorized receivers will likely be fried with voltages well under 100 volts.   There are a couple of solutions that can offer lower-voltage clamping more appropriate to these lower-power transistorized radios.

Certainly the grounding antenna switch described above, as well as some coax cable somewhere if possible, should be utilized. Shortwave radios with a collapsible antenna should have it at minimum size when not in use. The case of transistor-based equipment should ideally be constructed of metal, for its shielding ability, and ideally grounded, and any wires longer than a couple inches (e.g., Morse code key, speaker wire, headphone wire) should have protection added.   A typical electromagnetic-interference technique of adding a 0.01- 0.1 microfarad ceramic capacitors across those kinds of low-frequency signals wires (connected to ground/shield) will knock out a lot of the higher frequency components of an EMP E1 surge.

To provide further voltage clamping for those low-signal lines, a metal-oxide-varistor device ("MOV", Figure 4-4) can also be placed across speaker and similar wires (not across RF frequency wires).   These semiconductor devices do not have a "polarity" and do not conduct until their breakdown voltage is reached. They come in a dizzying array of protection (breakdown) voltages, and can protect repetitively unless the onslaught is extremely large.   An 18-volt MOV suitable for speakers, headphones, many Morse code keys, and probably digital data lines as well, is available for less than $1. [33]   One can select higher or lower voltage-ratings as appropriate.

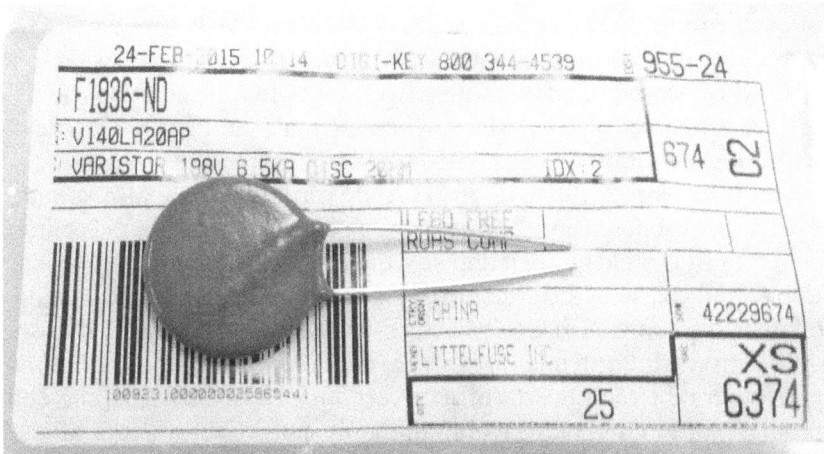

Figure 4-4: Metal oxide varistor designed to pass 140VAC (RMS) and to clamp at 198 volts, capable of shunting 6,500 amperes for 20 microseconds (which is comparable to an E3 event and much longer than an E1 EMP pulse).   These are available in many different breakdown voltage levels, including lower voltage levels suitable for Morse code keys, speaker and headphone wires.

To protect a transistorized receiver, "back-to-back" silicon diodes (see schematic, Figure 4-5) can be connected at the receiver input, between the center conductor and the shield, either connecting inside the feedline connector, or added inside the receiver's enclosure. The reverse breakdown voltage rating is of lesser importance. One diode conducts and limits the voltage to 0.6 V when the center conductor is driven positive, and the other conducts and limits the voltage to -0.6V when the center conductor is negative, so the peak-to-peak AC voltage is limited to 1.2 VAC. Because the transmitter may well produce more than this voltage, there has to be some relay or other mechanism to disconnect the receiver from the transmission line during transmitting, so these diodes don't short out the transmitter output. This type of protection has been used in some commercial and homebrew QRP transceivers to protect the receiver even from the transmitter output. [34]   A small signal diode such as a 1N914 or 1N4152 will have a negligible amount of capacitance of a few picofarads and a very fast response, but be able to carry a current of a few amperes for a microsecond. At the cost of a bit more parallel capacitance added to the line, a heavier power diode such as the 1N4004 or 1N4007 can be used for the back-to-back diodes, with a peak current capability of more than 30 amperes. For most of the high frequency ham bands, the added capacitance of the heavier power diode will not be noticeable to the receiver. When I tested this at 7 MHz, there was absolutely no discernible loss in received signal strength. However, in my heart of hearts, I don't trust just two diodes to protect a transistor receiver connected to a very long wire antenna from EMP. I would therefore use four diodes (2 small-signal diodes like 1N914, and 2 heavier ones like 1N4004/ 1N4007) AND further I would also connect a 60-volt gas discharge surge arrestor (see next section) from the center conductor to ground as "belt and suspenders" protection for my valuable radio! In the case of a transistorized shortwave antenna with a small collapsible antenna, using two back-to-back diodes connected with short wires between the antenna (at its base) and chassis "ground" or "ground" on the internal circuit board (often a very long trace encircling the entire border of the circuit board) would probably add a large amount of protection.

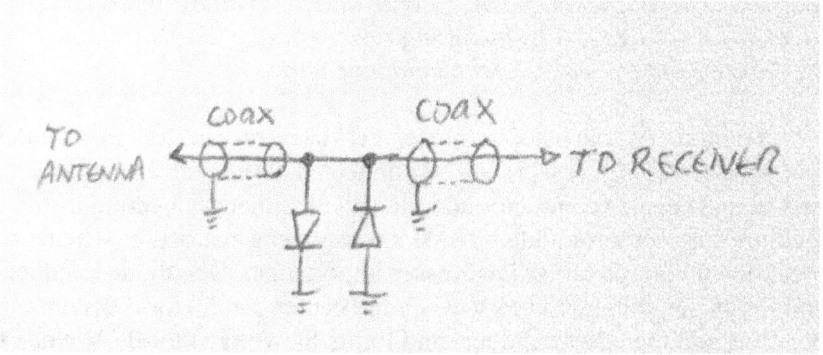

Figure 4-5. "Back-to-back" diodes shunting both positive and negative higher voltages to ground so that only very small voltages can reach the receiver.

Fully protecting the antenna output of a transistorized QRP transmitter is a bit more difficult because one must again allow passage of the normal power output voltages generated for commonly-encountered SWR ratios, yet short out voltages just higher than those to protect the transistor amplifier. If for a single instant, a power amp transistor or any other within the transmitter is exposed to, say, a voltage greater than its rated Vceo or Vcbo voltage, that transistor is likely to be irreparably damaged. (It might be a good idea to stock up on some spares.) While the fractions-of-a-second punch of lightning is long enough to literally melt/ burn conductors and thus leave an indelible mark wherever it coursed, the E1 EMP is gone so quickly that likely only the PN junctions of semiconductors are invisibly but very effectively destroyed. Some transistor designs have relatively little margin of voltage safety. However, it is also possible that the output tuned circuit of the amplifier will screen out some of the frequency spectrum from reaching the vulnerable transistor, also. If we assume an SWR of better than 2:1 and the common 5-watt output, the output voltage is on the order of 31VAC RMS, or under 50 volts peak. One immediately thinks of using a MOV (available in all manner of voltages) to provide protection. The problem with using metal oxide varistors is that they have far larger capacitance (often THOUSANDS of picofarads) than is tolerable in the matching circuits of typical HF transmitters. To get an acceptably lower capacitance requires once again using a gas-discharge tube, such as the Bourns Inc. 2020-15T-C2LF. This gas discharge tube has a breakdown at 60volts (the selection of breakdown voltages for gas-discharge is somewhat more limited than for MOVs), and can shunt up to 10 kiloamperes. It is manufactured in a 3-wire version with two surge arrestors connected together. Simply connect the center wire of the pack to the transmitter output, and both outer wires to the shield (ground). The capacitance is less than 1 pF; it will not affect your transmitter's tuning, and the price is less than $3! [35] It may possibly provide adequate protection for the transistorized QRP transmitter. However, this obviously cannot be guaranteed, as we are dealing with enormous instantaneous currents. To give your gas-discharge surge arrestor a fighting chance to clamp that voltage sufficiently for sensitive QRP rigs, I suggest that you add a 1-ohm carbon or carbon-film resistor in series with the center conductor, on the ANTENNA side of your 60 volt gas discharge surge arrestor. (See Figure 4-6) This can be a common 1/2 watt or 1 watt resistor, just NOT a wire-wound one. It will not appreciably affect your transmitted power.

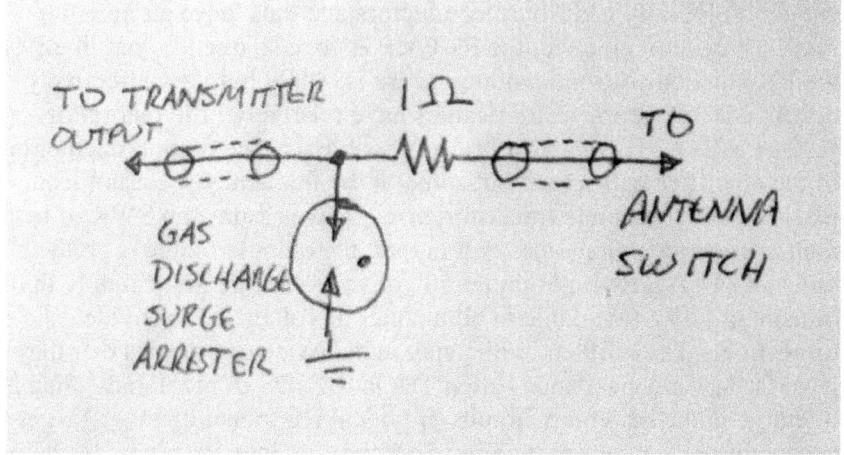

Figure 4-6. Combination of gas discharge surge arrestor and series 1 ohm resistor to protect 5W output transistor transmitter. Voltage to the transmitter may be limited to approximately 60 volts.

CB RADIOS: These same solutions just described for transistorized QRP transmitters above can be applied to common transistorized CB AM transceivers. 12-watt single sideband transceivers with feedline SWR below 2:1 may also work well with this system.

In conclusion, for about the price of a fast-food meal and a bit of wiring, one can purchase and install all the protection devices reasonably useful to protect either a QRP transistorized or higher-power vacuum tube or solid state ham radio station, and thus have a reasonable chance of surviving a first or second strike even while using the equipment. Equipment not in use should be disconnected from antennas; a shorting antenna switch can make this more convenient. Spare equipment should be left completely unconnected, and if economically possible, backup radios stored in a Faraday cage. As a side benefit, one gains some appreciable near-vicinity lightning protection. Powerline protection from E2 / E3 components of an EMP attack will be discussed in Chapter 5.

CHAPTER 5
HARDENING RADIO POWER SYSTEMS
AGAINST E3 / CME DAMAGE

Protecting HF/VHF communications gear against the E3 wave requires different (and easier) techniques than protecting against the broad-spectrum E1 component. Protecting against the E3 wave should also provide protection against a Carrington-level CME solar event.

While the E1 wave comes from energized electrons moving in the earth's magnetic field, the E3 wave's genesis is the forcible shoving aside of the earth's magnetic lines of force by the atomic/nuclear electromagnetic output. The time-scale for the E3 signal is much longer (many seconds), and the waves of much lower frequency. Rather than attacking equipment via their antennas, the E3 wave causes the (normally static) magnetic fields on the earth to become temporarily time-varying, and thus turns all very long wiring into an electrical generator for several seconds. The principal vulnerability are the power lines of the national electrical grid, which (with their ground return paths), encircle vast geometric areas through which this time varying flux can pass, generating very low frequency/quasi-DC currents (often called GIC: ground-induced-current) of very high magnitude (thousands of amperes) in long wires of the electrical grid. These extremely large quasi-DC currents will push the magnetic cores of grid-connection transformers into saturation, which causes increased losses that will damage these transformers, likely irreparably in many cases. For many 345kV-level transformers, replacements will not be available for months to years. *The grid goes down.*

Proposed solutions to block these quasi-DC currents in the vast interconnection lines of the national grid include placing large capacitors in series with the high voltage wiring to limit the DC current; another (cheaper) proposed solution is to place huge 5-ohm series resistors in the neutral-to-ground connection of these transformers, in hopes of limiting the induced current. Implementation of these solutions has been frustratingly slow.

It is not that difficult to protect HF communications gear from a solar flare or E3 EMP type disturbance. The key anomaly mentioned so far is damaging voltages which may come down the household power line because of the huge currents induced farther upstream in the power grid. An additional concern is that if neutral (center-tap) transformer wiring becomes damaged by the GIC currents at any level in the power transmission system, some neighborhood distribution systems may suddenly receive not just transiently higher voltage, but steady-state (continuous) significantly higher (or lower) voltage. The loss of a neutral wire in a transformer will cause downstream voltages to become unpredictable, as each load becomes part of a voltage divider network, instead of being constrained to the normal two legs of 120 volts each, with a total of 240 volts between the two "hot" wires fed to the household. The power supply of your communications equipment might be permanently destroyed by, say, a wall voltage that suddenly becomes 182 volts, instead of 115!

In order to protect against these risks, one needs first metal-oxide-varistor (MOV) surge protection across every pair of power wires to their equipment (hot to neutral, neutral to ground, ground to hot). While cheaper "surge protectors" may only have MOV protection across hot to neutral, higher quality surge protection devices usually have MOVs protecting all three combinations of the power lines. And it is easy to purchase your own 200V or 220V MOVs and wire them across each possible pairing of wires, as shown in the accompanying Figure 5-1. (The peak AC voltage is much higher than the 120V RMS voltage, and power line voltage can easily move up 10%; making the clamp voltage too low risks destroying the MOVs during relatively normal swings.)

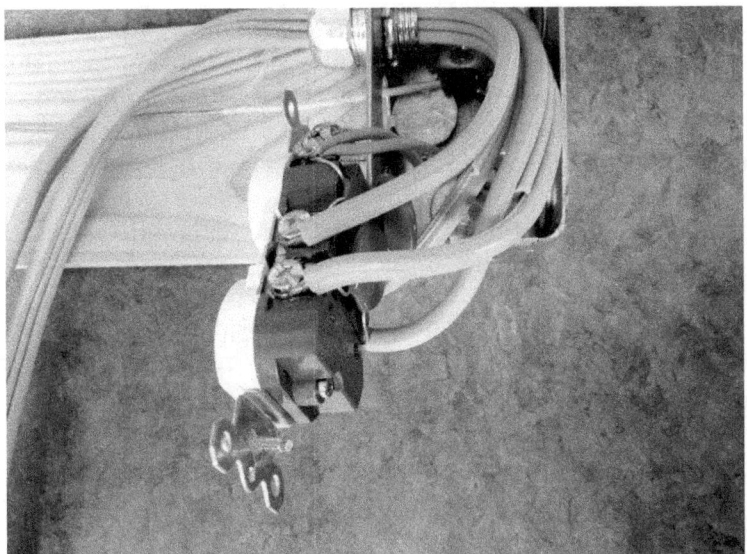

Figure 5-1 MOVs *(red circular disks) connected across all three combinations of hot/neutral/ground wires of an outlet prepared to protect communications equipment.*

Secondly, the equipment needs to be promptly removed from the commercial power mains in the event that the household voltage suddenly becomes far higher than normal, due to a neutral wire failure as a result of GIC. One way to assure this is to use a commercial uninterruptible power supply (UPS) which is designed to switch the equipment to battery-generated AC power in the event of either low or high voltage. One such device, which would temporarily power even a vacuum tube type ham radio, is APC BE750G 750VA uninterruptible power supply available from Amazon (and other vendors) at: http://www.amazon.com/APC-BE750G-Back-UPS-10-outlet-Uninterruptible/dp/B000Z80ICM/ref=sr_1_4?ie=UTF8&qid=1433778408&sr=8-4&keywords=%22apc+ups%22

More and more UPS backup-systems available to the consumer have this ability to quickly disconnect from AC line voltages that are out of the normal range. You can easily pick up a workable model for $50 or less.

To minimize disparate voltages in different equipment, try to establish a "single point ground" at your communications equipment. This can be as simple as a heavy-gauge length of bare wire running along the back edge of a desk. Connect that "ground" to the best true ground wire that you have (which might be the house ground wiring, or might be a shortest-path-possible wire out of your house to a long ground rod driven into the earth). Run a short, direct grounding wire from the chassis of every device that is part of your communications equipment to that "single point ground". Try to power every piece of your communications equipment from a single power circuit/outlet, ideally the uninterruptible power supply discussed above.

DC power supply lines (typically 13.8 VDC) can be additionally protected using Transient Voltage Suppressor (TVS) diodes, which are somewhat like zener diodes, and conduct heavily when the voltage exceeds a set amount. These are available in both unidirectional and bidirectional versions. The bidirectional versions can be easily wired across the 13.8VDC power wiring without regard to polarity.[36] An example of a bidirectional TVS diode that may work well on 13.8VDC power wiring at a cost of much less than $1.00 is the Vishay 1.5KE18CA.[37]

Finally, since either a Carrington type event or an EMP attack is likely to mean that commercial power is quickly lost after the initial voltage aberrations, and unlikely to return for months to years, one needs to have an alternate source of power, that is protected in such a way that it will remain viable after an EMP / Carrington event.

Generators should be kept COMPLETELY disconnected from all wiring possible -- wired-in permanent and automatically connecting generators may well be damaged by the E1, E2, or E3 components of an EMP. Many generators with a metal casing may provide some protection to their internal circuitry from E1 EMP waves. Many home generators include some electronic circuitry for autoregulation of the voltage, or may even have a built-in inverter. This circuitry may be damaged by the E1 EMP wave. Keeping a spare voltage regulator specific to your generator, stored in a Faraday shield, might be prudent. And as the generator may be needed for quite some time, a spare set of brushes, spark plug, etc., might be wise. Parts for your generator might be obtainable from the manufacturer; alternatively, a dealer whom I have found helpful to obtain parts for obscure generators (including Chinese ones) is http://www.generatorguru.com/

In conclusion, protecting against the E3 component of an EMP, or a

Carrington Event, is really rather simple, and most of it can be accomplished simply by purchasing an adequate uninterruptible power supply with quality surge protection, or a quality surge protector that has three-line protection.   Get your protection in place today!

CHAPTER 6
HARDENING ANTENNA SYSTEMS
AGAINST E1/E2

Bodson found that RG8 coax cable was very likely internally arcing during EMP simulation tests, limiting the damaging voltages delivered to radio circuitry to approximately 5 kilovolts. Assuming 50 ohms, this would represent 500,000 watts delivered (for a brief instant) to the radio. Using a simplifying (and admittedly over simplified) assumption of level power distribution from 0 to 100 MHz, this would result in 5 kilowatts peak energy per MHz bandwidth. Many HF transceivers' antenna impedance matching systems will have a bandwidth equal or greater than 1 MHz, indicating that 5 kilowatts might instantaneously assault whatever stage is connected to the antenna in the event of an E1 event. This translated to roughly 500 volts RF RMS (over a 1 MHz bandwidth) --- and that is plenty to fry any transistor or integrated circuit input.

As discussed earlier in this text, this damage can be ameliorated by employing gas-discharge tubes appropriately in the transmission line. In my own protection systems for transceivers of roughly 100 watts, I have utilized 230-250 Volt gas discharge tubes, making some allowance for SWRs in the range of 2:1 or 3:1 maximum.

However, the choice of antenna and/or matching network can also significantly add to the protection.

Keeping in mind the frequency distribution of EMP E1 power presented earlier, it is extremely important to limit the unnecessary low-frequency pass-through of E1 energy, while limiting the unnecessary higher-frequency response as well, to slow the risetime of the signal reaching the protection overvoltage devices. Most amateur HF systems have no need for transmission of RF below 3 MHz. 2-meter VHF systems don't need any RF passage at all below 140 MHz. If the unnecessary frequencies can be shorted out or mismatched, considerably lower amounts of damaging RF can reach the transceiver and/or gas-discharge protection systems.

Examples of systems with innate E1 hardening:

- 2-meter or 70 cm "slim jim" or J-pole antennas[38] --- the shorted matching section used to match the antenna at the desired frequency has the wonderful characteristic of basically shorting out HF frequencies, dramatically reducing the damaging signals passed through.

- T-network antenna tuners -- the most common antenna tuner sold by manufacturers such as MFJ for HF usage. These have an adjustable inductor as part of the matching network that connects to ground and would tend to short out signals of lower-than necessary frequencies.

- T- and gamma-matched systems for connecting coaxial cable to YAGI driven elements -- these provide a near-direct-short for much lower than design frequencies.[39]

Additional protection systems can be added to many antenna systems:

- A shorted quarter wave transmission line stub at the desired operating frequency appears to be an open circuit at that frequency and can be connected right at the antenna (or at the transceiver) from center conductor to ground, directly shorting out frequencies much lower than the desired operating frequency and is an excellent protection system to add to fixed band systems such as 2-meter and 70-cm repeaters. Use an antenna analyzer to correctly size the 1/4 wave stub as the velocity factor of the transmission line may not be what you expect.[40]

- For HF antennas, an RF choke of sufficient inductance to present say 400 ohms of inductive reactance at the lowest frequency of interest will have little impact when placed in parallel with a 50 ohm resonant antenna....but will significantly short out E1 components below 1 MHz.

Whenever not in use, it would be a wise idea to use a grounding switch to disconnect the antenna from the communications equipment.

CHAPTER 7
MAINTAINING ELECTRICAL POWER
DESPITE EMP/CME

*A prudent man foresees the difficulties ahead and prepares for them;
the simpleton goes blindly on and suffers the consequences.*
Proverbs 22:3

An EMP or CME event can be reliably expected to disrupt all normal sources of commercial AC power. If you are to continue providing communications, you will need alternate power sources that will have survived the EMP/CME event, and are also capable of surviving a later repeat EMP attack.

Power sources break down into batteries, generators (including automobile alternators) and solar panel systems.

Batteries are basically impervious to EMP/CME because of their rugged construction and low internal impedances. However, only a limited amount of power can be stored, so that the long-term usefulness of batteries for continued society usage of communications systems is limited, past the first several hours.

Generators would have had to be protected from E3 /CME events either by disconnection for significant lengths of power cabling, or good installation of protective Metal Oxide Varistors. It would be wise to maintain a stock of backup brushes and voltage regulators (if your generator uses a voltage regulator). These can sometimes be found easily from Internet resources such as www.generatorguru.com

Solar
The survival of solar panel systems from EMP is debatable. The solar panels themselves are semiconductor devices (a diode) and potentially may be destroyed by the extremely high voltages and currents that might exist on their wiring. However, good protection by metal oxide varistors placed physically CLOSE to the solar panels (to avoid

transmission line effects at higher frequencies of E1 incident wave) may possibly serve to protect these energy-producing semiconductors from damage. When choosing MOV protection devices for solar panels, remember that the solar panel produces much higher output voltages during colder weather, and also take into account the tolerance error of the MOV breakdown voltage, to avoid your MOV shorting out your solar panel during weather and device extremes.

Extreme efforts to protect individual panels might include low-pass filters, such as those described later in this chapter to reduce inverter interference, or homemade using series inductors and parallel capacitors to form a low-pass-filter that would allow DC power to be provided by the solar panel, but prevent AC currents higher than a few tens of kilohertz from reaching the solar panel. Such coils (mounted right at the output of individual panels) should be constructed of adequately sized conductors to pass the rated current of the solar panel. A reasonable homemade design would be a series inductor of 20 turns of #14 insulated THHN house wire around a 1/2" steel bolt core, with 0.1 microfarad 100 volt ceramic or tantalum capacitors to ground on both solar panel and charge controller ends of the inductor.

And of course, spare solar panels can be wrapped in aluminum foil or stored in well shielded metal structures as Faraday cages.

While grid-connected inverters are typically protected with surge arrestors for significant surges and can withstand up to 6kV surges[41], it may be wise to add additional protection at all external connections to wires longer than a foot or two. This would include appropriately chosen Metal Oxide Varistor (surge arrestors). Midnite Solar markets several commercially assembled systems that would assist in this goal. Adding low-pass filtering systems that would block the higher frequency components of an E1/E2 wave is also a good idea. Be certain to use systems with adequate current carrying capacity. The same type low pass filter recommended previously to reduce RFI from generators to HF gear can be utilized for this purpose also; it is rated for 20 A RMS AC, and should be capable of handling up to 20A DC as well: Schaffner FN2030A-20-06 Power Line Filter, rated 20A AC, purchased from Mouser Electronics, www.mouser.com, details at: http://www.mouser.com/ProductDetail/Schaffner/FN2030A-20-06/?qs= %2fha2pyFadujyPQkwXmbiEp4v1SwIQD%2faY%2fbAgajTvcbbNdiK %2fOAztg%3d%3d

Figure 7-1: *Power-line EMI filters installed within inverter output panel box to reduce broad-spectrum inverter-hash interference to communications equipment. The same style low-pass filters can be utilized to protect solar panels from E1 EMP.*

RF Interference Caused By Your Own Solar Inverters

While traditional generators will likely produce relatively little electromagnetic interference to HF communications gear, inverters or inverter-based generators may produce a broad-spectrum radio "hash" due to their fast switching transients, that will seriously degrade your HF equipment's ability to hear weak signals. This will be a significant problem during a crisis. In order to squelch that interference, a commercial filter can be added in the power output wiring: e.g., Schaffner FN2030A-20-06 Power Line Filter, rated 20Amperes AC, available from Mouser Electronics, www.mouser.com Details at:

 http://www.mouser.com/ProductDetail/Schaffner/FN2030A-20-06/?qs=%2fha2pyFadujyPQkwXmbiEp4v1SwIQD%2faY%2fbAgajTvcbbNdiK%2fOAztg%3d%3d

Comparable filter:

http://www.amazon.com/Power-Single-Phase-Filter-CW3-20A-

S/dp/B00D0U83D8

Such filters dramatically improved my ability to operate my ham radio station from my commercial solar power inverter.

CHAPTER 8
COMMUNICATIONS AFTER EMP

*We can make our plans,
but the final outcome is in Gods hands.
Proverbs 16:1*

Immediately after a high altitude EMP event, ionospheric high frequency communications will be halted by extreme ionization of the D layer, which absorbs skyward bound (and returning) shortwaves. Ground direct point-to-point communications, such as VHF/UHF, should continue to function normally, and many battery-powered small units without long antenna wires connected will survive and continue to operate normally. These can be very useful for local security patrol use, as well as neighborhood and even longer-distance communications and coordination.

My attempts to find out which high-price-tag commercial police/fire/emergency communications systems already include EMP protection have been frustrating. Here's my educated guess as to the outcome of an EMP/CME event (Table 8-1):

TABLE 8-1 Estimated commercial systems' survival

Communications system	EMP Event	CME event
Wired telephone systems	Destroyed	Destroyed
Trunked police/fire systems	Small handheld walkie talkies survive. Trunked systems (700/800 MHz) possibly survive E1. Loss of power from E3/CME eventually downs systems without longer term generator backup.	Small handheld walkie talkies survive. Trunked systems (700/800 MHz) possibly survive E1. Loss of power from E3/CME eventually downs systems without longer term generator backup.
Red Cross 47 MHz radios	Destroyed by E1.	Likely survive but require power.
Cell phone service	User phone survive. Cell tower radios small chance of survival but lose power quickly as backup power is only hours worth.	User phones survive. Cell tower radios undamaged but lose power quickly due to inadequate power backup.
Internet	Wired systems likely destroyed by E1; remaining systems lose power due to E3.	Primary risk is loss of power due to lack of backup.

Forestalling Anarchy As Long As Possible

With so many systems destroyed, the onset of anarchy will be rapid in all but the rare communities that posses both the most carefully thought-out radio protection systems, and the most disciplined and close-knit community architecture. The Amish come to mind.

For most communities, the provision of emergency communications by volunteers who have carefully protected their radio communications systems, will primarily have early value in providing situational awareness to rapidly-dissipating authorities and to others who have prepared for possible societal anarchy. It will forestall the onset of New Orleans Katrina-type anarchy for only a short time. After the onset of anarchy, small groups of well-prepared individuals would likely be the only groups with remaining communications capabilities.

Any remaining central government may declare a cessation of radio communications similar to what happened in some wars, but it will be difficult for them to broadcast those prohibitions, monitor their effects, or find people who still maintain communications.

Post-Anarchy Communications
For well-prepared individuals, communications goals will likely include:
- providing local hand-held communications abilities for security personnel (guards)
- longer range VHF or HF communications to maintain contact with allied and sister communities
- monitoring long-range HF communications for news that has survival benefit

Particularly for communications to allied and sister communities, encrypted or digital communications may have significantly increased importance, as well-prepared individuals and communities may not wish to advertise openly their capabilities and geographic locations. ***Unencrypted voice communications are a dead giveaway for radio direction-finding.*** Error-corrected packet technologies will allow encrypted (e.g.,PGP[42]) communications using public/private key technology and provide virtually impenetrable communications quickly enough that direction finding may be difficult for amateurs. Three possible strategies are likely:

- Using pre-arranged schedules for time and frequency, encrypted digital communications may be broadcast without ARQ error correction using such software as FLDIGI[43], on any frequencies suitable for the desired communication, including HF / VHF / UHF--requires excellent signal-to-noise ratio to succeed.
- Peer-to-peer encrypted attachments may be exchanged using WINLINK[44] peer-peer communications, which offer ARQ error-correction that will make encryption always successful.

- With beacons turned off, packet based technology using LINBPQ or WINLINK client-server technologies will allow non-scheduled rapid encrypted ARQ error-corrected transactions.

Clearly the well-prepared citizen will be aware of and able to scan for voice or digital un-encrypted transmissions of intelligence value during any period of anarchy. This will only be possible for those who practice these skills beforehand.

During the time period where geographic stealth must be maintained, directional antennas will lessen the radio energy escaping in undesirable directions. Clever usage of active and passive spectrum communications may play a role here as well.

Eventual Social Stabilization

It is not possible to predict how the social system will stabilize, but it is likely that groups with effective communications will be able to play a helpful role in arranging / advertising larger societal anchor events and arrangements such as:

- bazaars & swap meets similar to periodic marketplaces in developing societies
- elections as governmental structures re-emerge
- access to distant sources of more rare information, such as medical treatment advice and options

However, the real need for national survival is longer-ranged communications, such as microwave point to point, Internet, skywave shortwave. Many systems of those types will be destroyed by the E1 pulse, others by the E3 pulse, and still more when they lose power as their backup power system run out of fuel. Nevertheless, many individual shortwave transmitters will remain, principally well-planned ham radio stations and well-fortified government stations.

If you wish to maintain communications to loved ones or allies, you will possibly have considerable difficulty making contact with them if you do not have previously agreed upon times, and frequencies and calling strategies. Practice in this area is essential.

WINLINK

An alternative to hunting needle-in-the-haystack communications counter-parties is provided by the email-over-radio pioneered for the private and governmental sectors by the WINLINK group. Their Radio Message Server stations can provide automated radio-relayed messages from any part of the world to any other   Some of those stations, even in affected regions, may have had the foresight to install EMP protection and provide EMP-hardened backup power. If you are lucky enough to be able to contact one of the stations you may have a communications node that can be used over large distances to make needed communications, or to set up schedules to reach other counter-parties using direct communications.

EMP-Hardened Radio Communications

CHAPTER 9
CONCLUSION

He who is generous will be blessed,
For he gives some of his food to the poor.
proverbs 22:9

While the United States military (or that of other major nations) may possibly be somewhat prepared for an EMP attack or a CME event of Carrington-level magnitude, it seems clear that the United States population -- and likely that of many other nations -- is not at all.

Meanwhile, the list of nations who may well possess the nuclear capability to carry out such an attack (depending on the adequacy of our defensive missile detection and interdiction capabilities, or else the ability of an adversary to stealthily get a SCUD or other delivery device close enough to avoid those defenses) is growing considerably, and even non-state terrorists could mount such an attack.

The wise citizen would take to heart these risks and make all sorts of preparations for the anarchy that would likely follow either the naturally occurring CME event, or the successful EMP attack.

Maintaining the ability to receive and/or transmit radio information is likely an important part of such preparation. A checklist for such a capability would likely look like the following:

Radio (HF / VHF / UHF)
☐ Protected solid state or vacuum tube
☐ Gas-discharge tube feedline protection (either homebrew or commercial)
☐ Backup units protected in Faraday containers

EMP-Hardened Radio Communications

☐ AC line MOV Protection (trivial cost)

☐ AC line automated disconnect in the event of dangerous voltages (<$100 UPS)

☐ Backup EMP/CME protected AC power source (e.g., generator)

☐ Sufficient fuel if required

☐ MOV protection of any solar power systems (trivial to modest cost)

☐ Optional: low pass filtering of solar power systems (modest cost)

☐ Carefully chosen antenna tuners to minimize E1/E2 pass-thru (easily obtainable used on ebay)

☐ Carefully chosen antenna systems to minimize E1/E2 pass-thru (easily home-constructed)

☐ High-directivity antennas where possible (for VHF, easily home constructed)

☐ Voice capabilities (included in most equipment)

☐ Multiple types of digital capabilities (modest cost, using generally free software)

☐ Ability to minimize transmission on time (use carefully chosen digital modes)

☐ Optional: passive or active repeater systems

The more of these checkoffs achieved by key members of the population, the better able they will be to maintain effective communications during/after an EMP/CME event. For further information, the reader is suggested to review amateur-radio related postings on www.survivalblog.com, and www.thesurvivalistblog.net, among others.

APPENDIX:
CONSTRUCTING A FARADAY CAGE

As the reader has no doubt already surmised, there is an extensive published body of knowledge about EMP and EMP-protection, both in the lay- and scholarly press, developed in the 70 years since the world's first man-made nuclear explosion. It is perplexing why the civilian population is so ill-informed and ill-prepared.

A very large number of authors have addressed in print and video how to construct a Faraday cage to protect unused and important electronic equipment that would therefore be in working condition after an EMP attack. This might include radios, voltmeters, standing-wave-ratio meters, computers, alternators, ignition components for vehicles, and many other important devices.

My suggestions for building a protective container are simply as follows: Start with a new galvanized steel trash can of appropriate size. These are available inexpensively from fireplace ashcan size to 50-gallon trash disposal size. Being new, the finish will be relatively bright and conductive. Find one with a well-fitting lid, that grasps tightly and with very minimal gap if any.

The shielding of a Faraday cage rests on corollaries of Maxwell's Laws --- there cannot be an electrical field inside a perfect conductor (for it would cause an infinite current), and, RF currents flow only on the "skin" (of finite depth) of a perfect conductor.

Of course, a steel trash can is an approximation of a "perfect conductor."

Hence, there is no requirement at all that the trash can be "grounded" in any way (there is nothing in the protection that rests on any such connection) -- only that it have a well-conducting and continuous external surface.

For lower frequencies, the "skin" depth can be appreciable and the trash can is relatively thin. For this reason, I personally would not assume that the inner surface will not have at least some voltage in the event of a real EMP attack, at least for a few nanoseconds. I would therefore line the trashcan with corrugated cardboard so that none of my precious cargo actually touches the metal of the inner surface of the can.

Emanuelson has made a very important point that the protection of a Faraday cage does not have to be infinite --- it only has to knock the incident E1 wave down by a sufficient number of decibels (dB) such that the field strength that reaches the electronics within it is no longer of a damaging magnitude. His calculations suggest that a possible 80 dB of protection would likely be quite sufficient.[45] One may choose to "nest" items inside aluminum foil inside the trash can if one is unconvinced, but if properly constructed, the can itself may well suffice.

Others have made convincing proofs that circular gaps in the lid can allow some amount of UHF energy to seep into the can.[46] While the E1 EMP wave is much weaker in the UHF frequency range, I believe their suggestion to augment the lid seal with a conductive aluminum duct tape is a very good one. By this, I mean an actual tape intended to seal air conditioning duct work, which literally has an aluminum base, and an adhesive that is chosen because it does not dry out like the fiber "duck" tape's adhesive does. Make an overlapping circumference around the joint between the lid and the trash can. The aluminum tape does not have to be bonded electrically to the can; capacitive coupling between the metalized tape and the can will make a sufficient connection.

That's all there is to it. I recommend that you make a list of what is inside the can and post it on the lid, as it is very easy to forget what is stored there.

REFERENCES

[1]   Report of the Commission to Assess the Threat to the United States from Electromagnetic Pulse (EMP) Attack. Critical National Infrastructures. Figure 1-1, p 2.   Accessed at: http://www.empcommission.org/docs/A2473-EMP_Commission-7MB.pdf

[2]   R. James Woolsey and Peter Vincent Pry, The Growing Threat From an EMP Attack, Wall Street Journal, Aug 12, 2014.   Accessed at: http://www.wsj.com/articles/james-woolsey-and-peter-vincent-pry-the-growing-threat-from-an-emp-attack-1407885281

[3]   Pete Riley, On the Probability of Occurrence of Extreme Space Weather Events.   Presentation to the SSB Committee on Solar and Space Physics, October 7, 2014.   Accessed at: http://sites.nationalacademies.org/cs/groups/ssbsite/documents/webpage/ssb_153147.pdf

[4]   F. G. Gosling, United States Department of Energy, The Manhattan Project Making the Atomic Bomb, January 1999 Edition. 2010 version accessible at: http://energy.gov/management/downloads/gosling-manhattan-project-making-atomic-bomb

[5]   Jerry Emanuelson, An Introduction to Nuclear Electromagnetic Pulse, Futurescience, LLC www.futurescience.com/emp.html

[6]   General Loborev, director, Central Institute of Physics and Technology, June 1994 presentation; Overview presented by the Commission to Assess the Threat from High Altitude Electromagnetic Pulse (EMP) Accessed at: https://images.military.com/DT/images/Graham.pdf

[7]   Dennis Bodson W4PQF, Electromagnetic Pulse and the Radio Amateur. QST, August 1986, 15-20, 36.   Accessed at: http://www.arrl.org/files/file/Technology/tis/info/pdf/88615.pdf

[8]   Report of the Commission to Assess the Threat to the United States from Electromagnetic Pulse (EMP): Critical National Infrastructures, April 2008, pp. 42, 43. Accessed at: http://www.empcommission.org/docs/A2473-EMP_Commission-7MB.pdf

[9]   Report of the Commission to Assess the Threat to the United States from Electromagnetic Pulse (EMP) Attack. Critical National Infrastructures. Figure 2-3, p 33.   Accessed at: http://www.empcommission.org/docs/A2473-EMP_Commission-7MB.pdf

[10]   Near Miss: The Solar Superstorm of Juy 2012. NASA Science Beta. July 23, 2014.   Accessed at: https://science.nasa.gov/science-news/science-at-nasa/2014/23jul_superstorm

[11]   Report of the Commission to Assess the Threat to the United States from Electromagnetic Pulse (EMP) Attack. Volume 1: Executive Report 2004. Accessed at: http://www.empcommission.org/docs/empc_exec_rpt.pdf

REFERENCES

[12] Report of the Commission to Assess the Threat to the United States from Electromagnetic Pulse (EMP) Attack. Critical National Infrastructures. Accessed at: http://www.empcommission.org/docs/A2473-EMP_Commission-7MB.pdf

[13] 2006 North Korean nuclear test, Wikipedia. Accessed at: Accessed at: https://en.wikipedia.org/wiki/2006_North_Korean_nuclear_test

[14] Woolsey RJ, Pry PV. The Growing Threat from an EMP Attack, WSJ Aug 12, 2014. Accessed at: http://www.wsj.com/articles/james-woolsey-and-peter-vincent-pry-the-growing-threat-from-an-emp-attack-1407885281

[15] Agence-France-Presse, NORAD Moving Comms Gear Back to Mountain Bunker, Defense News, April 8, 2015. Accessed at: http://www.defensenews.com/story/defense/international/americas/2015/04/08/norad-moving-comms-gear-back-to-mountain-bunker/25470435/

[16] EMP Pulse Generators, Information Unlimited, Accessed at: http://www.amazing1.com/emp.html

[17] Jenni Bergal, States Work to Protect Electric Grid, PEW Charitable Trusts, Feb. 27, 2015. Accessed at: http://www.pewtrusts.org/en/research-and-analysis/blogs/stateline/2015/2/27/states-work-to-protect-electric-grid

[18] Saraswat, Trends in Integrated Circuits Technology, EE311 Notes, Accessed at: http://web.stanford.edu/class/ee311/NOTES/Trends.pdf

[19] Report of the Commission to Assess the Threat to the United States from Electromagnetic Pulse (EMP) Attack. Volume 1: Executive Report 2004. Figure 2, page 6. Accessed at: http://www.empcommission.org/docs/empc_exec_rpt.pdf

[20] Edward Savage, James Gilbert, William Radasky. The Early-Time (E1) High Altitude Electromagnetic Pulse (HEMP) and Its Impact on the U.S. Power Grid. Metatech. Report prepared for Oak Ridge National Laboratory. January 2010, taken from table 2-1. Accessed at: https://www.ferc.gov/industries/electric/indus-act/reliability/cybersecurity/ferc_meta-r-320.pdf

[21] Edward Savage, James Gilbert, William Radasky. The Early-Time (E1) High Altitude Electromagnetic Pulse (HEMP) and Its Impact on the U.S. Power Grid. Metatech. Report prepared for Oak Ridge National Laboratory. January 2010, Accessed at: http://web.ornl.gov/sci/ees/etsd/pes/pubs/ferc_Meta-R-320.pdf

[22] Jerry Emanuelson, *EMP Myths*, section on "either-or" myth. Accessed at: http://www.futurescience.com/emp/EMP-myths.html

[23] Jerry Emanuelson, *EMP Myths*, section *Myth:* Multiple EMP detonations can be used in a single attack to enhance E1 Accessed at: http://www.futurescience.com/emp/EMP-myths.html